YOUR KNOWLEDGE

- We will publish your bachelor's and master's thesis, essays and papers

- Your own eBook and book - sold worldwide in all relevant shops

- Earn money with each sale

Upload your text at www.GRIN.com and publish for free

GRIN

The Limits of Europeanisation. Domestic Constraints on Ownership Unbundling of National Electricity Systems

Nico Miguel

Bibliographic information published by the German National Library:

The German National Library lists this publication in the National Bibliography; detailed bibliographic data are available on the Internet at http://dnb.dnb.de.

ISBN: 9783346203694
This book is also available as an ebook.

© GRIN Publishing GmbH
Nymphenburger Straße 86
80636 München

Print and binding: Books on Demand GmbH, Norderstedt, Germany
Printed on acid-free paper from responsible sources.

The present work has been carefully prepared. Nevertheless, authors and publishers do not incur liability for the correctness of information, notes, links and advice as well as any printing errors.

GRIN web shop: https://www.grin.com/document/908564

Master Thesis

MA European Studies: Public Policy and Administration

The Limits of Europeanisation: Domestic Constraints on Ownership Unbundling of National Electricity Systems

Nico João Miguel

MA Thesis

30 June 2020

Abstract

Combined ownership of generation, transmission and distribution facilities enables vertically integrated electricity utilities (VIUs) to curb competition by means of various discriminatory practices. Therefore, the gradual unbundling of VIU's ownership structures constitutes a main approach of the European Commission towards completing the internal market of electricity. Following the third electricity directive, member states could choose between 'hard' unbundling regimes (combined ownership prohibited or highly restricted) and a 'soft' regime (separation of organisational functions but combined ownership allowed). This master thesis argues that a government's ability to choose a hard model is compromised by internal governance and interest structures which determine the individual costs of adaptation. Using an explanatory mixed-method design, I find that ownership of VIUs by sub-state veto players, cost-based network pricing regulation and private ownership of VIUs increase a country's probability to maintain a softer unbundling regime. The findings demonstrate how the pace and timing of the Europeanisation of electricity systems is restrained by nation states' reform capacities and emphasise the particularly constraining role of federalism.

Table of Contents

Introduction 4

I.   Unbundling as Key Ingredient for the Internal Market of Electricity 9

    Unbundling Options after the Third Electricity Directive 9

    Costs and Benefits of Ownership Unbundling 11

II.   Theoretical Framework 16

    The Goodness-of-Fit Hypothesis 16

    Rational-Choice Institutionalism 17

    Path Dependence 19

III. Research Design 20

    Variables and Hypotheses: The Determinants of Unbundling Regime Choice 20

    Methodology 23

IV.   Quantitative Analysis 25

    Decision-Theoretic Model 27

V.   Qualitative Analysis 31

    Case Selection 31

    Case Study Germany 32

    Case Study Netherlands 36

Conclusion 42

References 45

Figures and Tables

Figure 1 Discriminatory potential by unbundling model 10

Figure 2 TSO-unbundling models by country 11

Figure 3. Decision-theoretic model 28

Figure 4 Predictions vs observations decision-theoretic model 30

Figure 5 Revenue development DSOs Netherlands 2000-2008 41

Figure 6 SAIDI Index-Netherlands 2000-2007 41

Table 1. TSO Unbundling Regimes after the Third Directive 10

Table 2. Benefit-cost Analysis of Ownership Unbundling 13

Table 3. Determinants of unbundling regime choice 22

Table 4. Fisher's exact test of unbundling regime choice 25

Table 5 Sub-game results of decision-theoretic model 29

Table 6 Fisher's exact test of decision-theoretic model 30

Table 7. Case selection 31

Introduction

Reliable energy supply is the lifeblood of an economy. Therefore, it does not come as a surprise that cooperation in this domain has been a cornerstone of the European integration process from the onset. The creation of the European Coal and Steel Community (ECSC) and later the Euratom improved nation-states' access to primary energy sources and thereby fostered peaceful cooperation between them even beyond this sector. However, the internal workings of member states' electricity supply infrastructures remained untouched for a long time. By conventional wisdom, network industries were thought to be immune to competition and hence countries were divided into regional service zones where a single utility controlled the whole supply chain (Brunekreeft & Meyer, 2009). This paradigm changed when experts concluded that the prevailing system was inefficient and unsustainable. In the 1980s, in the context of the Single European Act (SEA), the European Commission raised in a working paper that aggregate welfare would substantially be increased, if member states allowed competition and market integration in the electricity sector. In the 1990s, these sentiments were translated into a concrete proposal for a liberalisation directive, and ever since national electricity sectors have undergone a constant reform-process, are deeply interconnected and, in most countries, customers have the freedom to choose their electricity suppliers (Dutton, 2015; Levi-Faur, 2004).

The recipe chosen by the Commission to promote competition was the unbundling of vertically integrated utilities' (VIUs) functions and structures. Vertical integration refers to the combined ownership of all stages of value creation - generation, transmission, supply, and distribution[1]. Holding companies across the whole supply chain enables VIUs to create barriers to entry for new players by various discriminatory practices. A critical aspect in this connection constitutes the ownership of networks. Both, transmission- and distribution networks are, in fact, natural monopolies due to long-term concessions and extraordinarily high economies of scale (Diekmann, Leprich & Ziesing, 2007). Thus, competition in these sectors is virtually impossible to take off as opposed to the stages of production and supply. As suppliers and producers rely on transmission and distribution lines to bring electricity to the consumer, the existence of a VIU would make them dependent on a direct competitor. VIUs can take advantage of this dependency in two important ways: First, they would be able to discriminate competitors' grid access by charging high network fees. Second, VIUs might cross-subsidise their supply and generation units by using the earnings generated in the monopolistic segment or vice-versa. Together, this may lead to the withholding of necessary investments in network infrastructures

---

[1] Generation refers to electricity production, transmission describes high-voltage networks, supply refers to the stage of selling electricity to the customer and distribution to the low-voltage networks connected to the households

and undermine competition (Brunekreeft & Meyer, 2009; Pollitt, 2007). Given this situation, separation appears to be an essential component of a truly liberalised market.

To account for adaptation problems, EU legislation follows a piecemeal approach and each new directive introduces a stricter minimum level of unbundling. The first liberalisation directive from 1996 only required *accounting unbundling* where the bookkeeping of the different units must be independent. The rationale was to combat the issue of cross-subsidisation and prevent the allocation of costs from supply or generation units to the networks and vice-versa (Brunekreeft & Meyer, 2009). Unsatisfied with the success of this model, the Commission soon drafted the 'acceleration directive' which entered into force in 2003. Henceforth, the new minimum level was *legal unbundling* (LU) where transmission system operators (TSOs) were to be held by independent legal entities. LU also involved *management unbundling* where management staff may not be shared as well as *informational unbundling* which forces companies to set up Chinese walls and fire walls to undermine the transmission of sensible data. While LU appeared as a big step forward and allowed a certain degree of competition, VIUs' ownership structures remained intact. As such, investment decisions were still made by the VIU as a whole. The strategic withholding of investment was therefore a main discriminatory tool under LU, leading VIUs to not invest in cross-country interconnections in order to isolate their home markets from foreign competition (Brunekreeft & Meyer, 2009).

The investment withholding by VIUs was a main argument in the European Commission's sector inquiry in 2007 to explain the sector's competition malaise (European Commission, 2007). Therefore, the following draft for a Third Electricity Directive sought to introduce full *ownership unbundling* (OU) of transmission system operators or an *Independent System Operator (ISO)*. Under ISO, the vertically integrated transmission operator is separated into two fully independent companies, where one owns the assets and the other one operates the grid. Investment decisions are exclusively taken by the operating company which largely prevents investors from using anti-competitive practices. As such, ISO has similar effects as a fully ownership unbundled regime. However, as both, OU and ISO went too far for several member states, after long-lasting negotiations, the final compromise introduced the *Independent Transmission Operator-model* (ITO) as a 'Third Way' (Brunekreeft & Meyer, 2009). The model is almost identical to LU with the main difference being that external representation must be separated and that sharing of staff is now entirely prohibited (Bundesnetzagentur, 2013).

Despite ample evidence that a hard regime (OU or ISO) is most beneficial in terms of competition (Van Koten & Ortmann, 2008; Pollitt 2007), several states still opted for the ITO model. The literature seeking to explain this variation is relatively scarce and mainly from scholars studying economics. Focusing on the second directive, Van Koten and Ortmann (2008) find a

5

correlation between perceived corruption in a country and unbundling regime choice. Their findings imply that in some countries incumbent utilities are more successful in their rent-seeking activities than in others. In a theoretical model, Lindemann (2015) highlights the impact of independent regulatory agencies on unbundling regime choice. He finds that if a regulatory agency is more loyal to the energy consuming industry rather than the energy producers, a government prefers full ownership unbundling.

More recently, Meletiou, Cambini & Masera (2018) show that majority private ownership and network pricing schemes are key factors explaining variation in unbundling regimes across countries in the EU. Pricing schemes are used to limit the network operators' ability to charge monopoly access fees by mimicking a market environment. Broadly, one can distinguish between incentive-based approaches and cost-based approaches. Cost-based regimes grant network operators a fixed rate of return on top of their regular costs. This ensures the profitability of the business. Under incentive regulation, the regulator sets a price- or revenue-cap and thereby incentivises the network operator to operate efficiently. The efficiency requirement of incentive regulation may make network ownership less profitable and lead VIUs to voluntarily give up their networks. Meletiou et al (2018) show that incentive regulation is positively associated with ownership unbundling. Regarding ownership, they imply that VIUs that were mainly owned by private owners are difficult to unbundle as the shareholders would seek to prevent forced expropriation of their assets.

Somewhat surprisingly, while institutional design choice constitutes a core problem in political science and public policy research, to date, no study in this field has addressed the drivers of electricity unbundling regime choice. There are several papers addressing the wider phenomenon of electricity sector transformation in the EU. For example, some scholars explain variation in liberalisation outcomes by pointing to the differences in nation states' preferences regarding the matter (Jordana, Levi-Faur & Puig, 2006; Levi-Faur, 2004). Focusing on the role of ideas, Bartle (2002) argues that institutions serve as a moderating channel for a general shift in society's preferences towards liberalisation and deregulation of network sectors. Thus, while according to this logic full liberalisation will be inevitable in the long-term, the pace and timing of this process is compromised by institutional factors. Taken together, these findings suggest that a focus on national institutional constraints is a promising avenue to understand member states' variation in unbundling regimes.

However, while Bartle (2002) stresses that national institutions may protract the introduction of electricity reform, his research is on a more general level and does not explicitly focus on the costs of adaptation resulting from compensating losers of reform and convincing opposing political veto players. By doing so, important nuances of national responses to exogenous reform pressures may be overlooked. This thesis addresses this gap by asking the following research question: What explains the variation in unbundling regime choice between European countries after the Third Electricity Directive? The example of ownership unbundling provides a relevant empirical example to unequivocally comprehend the redistribution of resources resulting from a liberalisation reform initiative. When governments plan to unbundle VIUs, they directly interfere with the interests of several private and public actors. Thus, depending on the exact interest and governance configurations, full unbundling induces varying costs of compensation and (re-)negotiation for governments. I argue that an adaptation-cost-based account of member states' reform capacities allows us to obtain a better understanding of the pace and timing of electricity sector liberalisation in the EU as well as processes of Europeanisation[2] more generally.

The research question will be addressed through the lens of the rational-choice variant of Börzel & Risse's (2003) goodness-of-fit framework which is complemented by the historical institutionalist notion of path dependencies. Based on the theoretical framework, I identify three domestic institutional constraints that determine member states' capacity to adopt ownership unbundling: (1) the existence of internal veto players, (2) the domestic ownership structures of VIUs prior to regulation and (3) transmission network price regulation. For the first constraint, internal veto players, two hypotheses are developed: First, it is argued that ownership of lower-tier government with veto powers decreases the likelihood for adopting a full unbundling regime. Second, I hypothesise that ideological incongruence across national legislative branches acts as an impediment to full unbundling. The second and the third constraint build on the findings by Meletiou et al (2018) but differ in important ways. It will be argued that not majority private ownership, but the mere existence of private ownership decreases the likelihood of a state to adopt a 'hard unbundling regime'. On a theoretical level, this variable is regarded through the lens of path dependencies highlighting that certain states may have locked themselves in the decision to privatise shares of their VIUs prior to liberalisation. As full unbundling would equal forced expropriation of private actors, it is much more difficult for these states to reach this level of liberalisation. Regarding transmission network pricing schemes, I argue that the mere presence of guaranteed profits by means of cost-based regulation acts as a constraint since network operators are less willing to give up this stable source of income.

To test these claims, this thesis uses an explanatory mixed-method design. In the first step, a quantitative analysis of 28 European countries in the form of a Fisher's exact test will be conducted, followed by a decision-theoretic model based on the significant variables. The second

---

[2] Electricity liberalisation is in this thesis treated interchangeably to Europeanisation as in the present case the impetus for national reform comes from the EU in the form of a directive

step substantiates the findings by a comparative qualitative analysis between the Netherlands and Germany, two similar countries with fundamentally different outcomes regarding this issue. The thesis makes a theoretical and an empirical contribution to the academic literature. On the theoretical level, the findings show how electricity liberalisation and Europeanisation are restrained by nation states' internal workings and emphasises the particularly restraining role of federalism. Empirically, the findings advance our understanding of unbundling regime choice, in particular by proving the influence of sub-state actors which has been disregarded in previous studies. Furthermore, the decision-model developed in this thesis can also be tested against the observations of other states outside Europe.

The first chapter presents the different models of unbundling after the third electricity directive and analyses the potential costs and benefits resulting from ownership unbundling from a social welfare perspective. The second chapter introduces the theoretical framework of this thesis. More precisely, the rational-choice model of the goodness-of fit hypothesis is presented, followed by an elaboration of the historical institutionalist logic of path dependencies. The next chapter presents the variables and hypotheses to be tested as well as the methodology. The fourth chapter and fifth chapters present the quantitative and qualitative findings. The last chapter summarises the findings and concludes.

I. Unbundling as Key Ingredient for the Internal Market of Electricity

Unbundling Options after the Third Electricity Directive

Following its 'energy sector inquiry' in 2007, the European Commission concluded that the continuing lack of competition and stagnant levels of cross-border interconnections are the result of persistently high degrees of vertical integration in national electricity systems. Consequently, Commissioner for competition policy Neelie Kroes pushed ownership unbundling in the centre of the legislative debate for a third directive (Meletiou, Cambini & Masera, 2018; Brunekreeft & Meyer, 2009). While the member states agreed on most other points of the legislative draft, an alliance led by France and Germany vigorously rejected the Commission's plans to fully separate VIUs. As such, the final compromise left member states with three options, which can be classified into a 'hard' and a 'soft' unbundling regime (see Figure 1 as a heuristic device). The following section presents these options and discusses how they differ in their potential to curb competition in the internal market:

Ownership Unbundling (OU)

This is the most rigorous form of disintegration and the preferred model by the Commission as articulated in Article 11 of Directive 2009/72/EC (European Parliament & Council of the European Union, 2009). Ownership unbundling requires companies along the value chain to be fully separate organisations with no associations in terms of ownership. This model makes formal strategic coordination practically impossible and is thus most conducive to competition (see Figure 1 on page 3). Within the third directive, OU mainly concerns the separation of transmission system operators (TSOs) from generation and distribution units as transmission ownership bears the highest potential for discrimination. However, the same logic can be applied to low-voltage distribution system operators (DSOs) (Brunekreeft & Meyer, 2009). Currently, the Netherlands is the only European country using full DSO-unbundling.

Independent System Operator (ISO)

The ISO model is a slightly attenuated form of ownership unbundling but can still be classified as a hard regime. Transmission operators can theoretically be split into two segments: a system operator (SO) managing the grid and a transmission owner (TO), owning the resources. The ISO model requires the system operator to be ownership unbundled from the system while VIUs are allowed maintain their assets. The independence of the operators prevents shareholders of VIUs to formally exert power on investment decisions and grid access conditions and thereby reduces the potential for discriminatory practices (see Figure 1) (Pollitt, 2007; Brunekreeft &

Meyer, 2009). A small number of countries in the EU currently have an independent system operator, including Ireland, the UK and Latvia (see table 1 on page 3). Italy, Greece and Hungary abandoned the model in favour of either full ownership unbundling or the less strict ITO model, which is explained below (Meletiou et al., 2018).

Independent Transmission Operator (ITO)

Giving in to German and French demands, the EU adopted the ITO model as a 'third way' for member states to retain the ownership structures of integrated utilities, provided they existed already prior to the directive. In most aspects, the ITO model is identical to '*legal unbundling' (LU)*, where transmission, generation and distribution units must be legally separated companies but can be held by the same investors. The main novelties of ITO are the prohibition of shared personnel and independent external representation. Thus, the model did not significantly decrease the anti-competitive potential of LU and can be classified as a soft form of disintegration (see figure 1). The model is currently used by several European countries, including Germany, France and Austria (Bundesnetzagentur, 2013; Brunekreeft & Meyer, 2009).

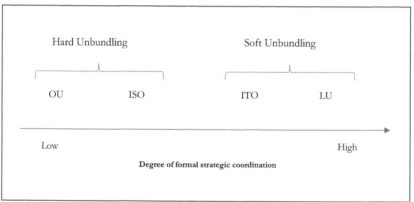

Figure 1 Discriminatory potential by unbundling model *(OU = Ownership Unbundling; ISO = Independent System Operator; ITO = Independent Transmission Operator; LU = Legal Unbundling)* (own compilation)

Table 1. Unbundling Regimes after the Third Directive (own compilation based on Meletiou et al (2018))

TSO Unbundling Regime after Third Directive	Countries
ITO	AT, GR, HU, HR, FR, BG, CH
MIX	UK (OU & ISO) , DE (ITO & OU)
ISO	LV, IE
OU	CZ,DK,FI ,IT, NL,NO,PL,PT, SK,SI, ES, SE, AL, AT, LT, RO, EE, BE, DE, AT

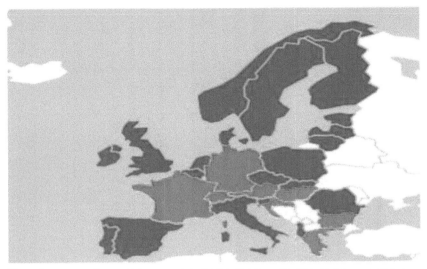

Figure 2 *TSO-unbundling models by country (Red = 'Soft Unbundling'; Green = 'Hard Unbundling')* (own compilation)

The comparison of member states' levels of implementation (Figure 2) reveals that, while the majority opted for a 'hard' regime, there is a 'red belt' across central Europe where countries still adhere to the softer ITO model. The aim of this thesis is to determine the main factors that explain why certain governments are so reluctant to 'Europeanise' their electricity systems while others have done it long before the EU pressured them to do so. The next part provides an overview of potential costs and benefits of a hard-unbundling regime that may have entered the calculations of governments when implementing the EU provisions.

Costs and Benefits of Ownership Unbundling

While the previous part has shown that the three available schemes provide utilities with different opportunities to curb competition, the following section will elaborate on the social welfare implications of full ownership unbundling. In a purely rational world, governments base their decisions on carefully weighed benefit-cost calculations. This is also applicable to the electricity sector which is marked by high resistance to reform and transformation. Various reasons may explain the rigidity of the network sector. First, a government needs to ensure security of supply at any time, considering factors that are beyond its own control. Second, as the sector is very capital intensive, reform measures must make sure to protect 'sunk' investments. Third, reforms must consider the physical realities that govern the electricity sector, (Pelkmans & Lusetta, 2013). All these factors have a crucial impact on the challenges associated with unbundling of VIUs. Various scholars comprehensively mapped out the costs and benefits of unbundling from a social welfare perspective.

Pollitt (2007) finds that empirically the most competitive electricity markets involve full ownership unbundling of transmission companies. Other scholars found that full unbundling of transmission networks is positively associated with openness to competition in the retail market, improved access to transmission and distribution for third parties, a more sophisticated wholesale trading system and much improved network congestion management (Copenhagen Economics, 2005; Joskow and Tirole, 2000). Copenhagen Economics' (2005) study also shows that electricity prices in the EU fell because of unbundling, controlling for other factors such as costs associated with the transition to green electricity. This resonates with the findings by Ernst & Young (2006) which also identified price decreases in the gas sector after unbundling. Alesina, Ardagna, Nicoletti & Schiantarelli (2005) provide empirical evidence of an increase in investment in the electricity sector following the introduction of unbundling. Zachmann (2007) shows that countries with a softer unbundling regime are less responsive to short-run cost fluctuations. Brunekreeft (2008) shows that generation capacity is greatly increased by ownership unbundling even though the effect is more accentuated for investment in internal networks and less in cross-country interconnections. All these studies support the argument that full unbundling indeed provides the incentives necessary to induce more competition in the electricity market.

However, there is also a wide array of theoretical and empirical arguments against full unbundling. Michaels (2006) argues that costs of capital may rise because of unbundling as incumbents profit from economies of scale as well as the loss of synergies along the supply chain. Focusing on transaction costs, Mulder & Shestalova (2006) argue that unbundling involves high costs associated with reorganisation and renegotiation. This is consistent with Pollitt's argument that unbundling should be accompanied by other far-reaching reform measures so that the necessary regulatory structures can be coordinated to the highest extent possible. Some studies find that legal unbundling brings about the benefits expected from ownership unbundling, if competition is already in place (Pollitt, 2007; Cremer, 2006; Bolle & Breitmoser, 2006). In a theoretical model, Cremer (2006) shows that legal unbundling might provide the welfare-maximizing optimal investment point. However, his model lacks empirical backing. Bolle & Breitmoser (2006) advance the claim that legal unbundling is more efficient than ownership unbundling as it prevents double marginalisation. This could be an issue if there is a quasi-monopoly or oligopoly in the producing sector paired with a monopoly in transmission, giving both the capacity to charge monopoly prices. It is evident that unbundling is a very complex undertaking. The table below from Pollitt (2007) presents a systematic overview of potential benefits and costs of full ownership unbundling.

12

Table 2. Benefit-cost Analysis of Ownership Unbundling (Politt, 2007, p.7-9)

Type of benefit/cost	Benefit	Cost
Effect on competition	Reduces scope for discrimination against non-integrated rivals.	May facilitate further generation mergers as sales of vertically unbundled transmission assets provide financial resources for horizontal integration.
Ease & and effectiveness on competition	Improves cost (and other types of) transparency in network and competitive businesses.	May increase requirement for regulatory oversight of transactions between unbundled stages of production.
Facilitation of privatisation	May make privatisation of competitive and network businesses easier due to sustainability (and hence reduced regulatory risk) of unbundled market structure.	May delay privatisation of network businesses because these can be retained in public ownership while generation and retail assets are privatised.
Security of supply	May improve transmission company focus on security of supply and incentivise improved information systems.	May create information problems between generators (electricity)/shippers (gas) and transmitters in the absence of investment in better information systems.
Transaction costs of unbundling	May reduce transaction costs by facilitating creation of more efficient price signals.	May increase costs if new computer systems needed to coordinate transmission and other separated segments. There may be also significant power renegotiation costs, which if with foreign parties may involve substantial wealth transfers and lower national social welfare.

13

Costs of capital/ investment	Overall cost of capital may decline if transmission business can get access to cheaper capital and if there is increased ease of integration of generation and retail. In an efficient capital market separation will lead to efficient cost of capital for each business.	May increase cost of capital and reduce investment if size of firm falls, or if regulatory risk is increased due to increased (and inefficient) regulatory oversight of investment decisions.
Synergy/Focus effects	Management of both parts of company may be subject to clearer incentives to improve business.	Loss of synergy (vertical economies) benefits due to smaller size or loss of experience of operation of other segments.
Double marginalisation	Not a problem when multipart-tariffs are in use.	May be an issue if available two part tariffs are not fully efficient.
Foreign takeovers more likely	Sale of assets may make efficient foreign (and domestic) takeover more likely. Undesirable takeovers of strategic assets may be covered by competition policy.	Sale of assets may lead to 'strategic' assets passing to foreigners if competition policy allows this.
Reduced risk of arbitrary government intervention	Unbundling likely to reduce government willingness (and need) to undertake further major reform for a period.	Unbundling may increase government interference in the operation of the network companies if these are kept in state ownership.

Pollitt (2007) weighs the costs and benefits outlined in the table above based on their theoretical likelihood as well as on insights from empirical studies and concludes that benefits clearly outweigh the costs. The author asserts that the effect on competition, ease of regulation, focus effects, privatisation, foreign takeovers and government intervention is likely to be positive based on the weakness of the counterarguments. For example, the effect of non-discrimination on competition is likely to have a stronger impact than the emergence of mergers of generation companies. Similarly, improved business incentives may outweigh the social cost resulting from the loss of synergies from separating vertically integrated utilities (Pollitt, 2007). It should be noted that, although the increased likelihood of foreign takeover is beneficial from a welfare-maximising perspective, from a political perspective it may be considered a cost. The third electricity directive in the EU has a provision that makes foreign takeover of networks more difficult (Brunekreeft & Meyer, 2009). As such, the EU effectively ruled this out as a decision criterion, although this may perhaps constitute a conflict with existing provisions in the Energy Charter.

On the cost-side, Pollitt's assessment shows that high transaction costs can be a strong disincentive to unbundling as the establishment of new computer systems and contract renegotiation costs are likely to be high. He remarks that the renegotiation costs highly depend on the contract parties. As such, he implies that unbundling of public utilities involves lower transaction costs than separating a private company. Furthermore, the creation of competition authorities and other necessary structures are likely to incur high upfront costs. (Pollitt, 2007). Moreover, Pollitt (2007) is not entirely certain about the effects on investment of ownership unbundling and subsequent evidence shows that in the short-term may even reduce investment due to the loss of scale economies (Gugler, Rammerstorfer & Schmidt, 2013).

While ownership unbundling is likely to bring about benefits in the long term, Pollit's (2007) assessment implies that high short-term transaction costs as well as insecurities over network investments may be a reason why governments are reluctant to adopt a hard regime in the short-term. The focus on transaction costs as a predictor of institutional change constitutes a foundational notion of rational choice-based approaches in the political sciences. As such, a transaction cost-based framework may be useful to answer the research question of this thesis. The following section introduces two major theories in the Europeanisation literature and integrates them into one common approach to explain the phenomenon of the differentiated Europeanisation-pathways of unbundling regimes across European countries.

II.   Theoretical Framework

The Goodness-of-Fit Hypothesis

The Europeanisation literature provides various approaches and concepts that help explain governmental choice in the implementation of a directive. A central hypothesis is the 'goodness of fit' hypothesis. Broadly, it argues that implementation of and compliance with a piece of legislation from the EU depends on how well this given law integrates with existing institutional and political structures of the jurisdiction (Mastenbroek & Kaeding, 2006). While the hypothesis can be addressed from various theoretical angles, they have several fundamental concepts in common, which in the following will be addressed.

Until the late 1990s, research on Europeanisation mainly followed a bottom-up perspective, focusing on the processes and political dynamics of delegation from the national to the supranational level. The emergence of the 'goodness of fit' hypotheses introduced a top-down perspective, focusing on the way how and to what extent supranational policies reconfigure domestic institutional structures. This perspective allows to explore explanations for differential integration pathways such as opt-outs from several member states from the monetary union, non-compliance with existing provisions or protracted implementation (Börzel & Risse, 2013).

The central tenet of all approaches is the notion of *policy misfit* as a condition for institutional change. As the logic goes, the higher the misfit, the higher the adaptational pressures of the national polity and thus the higher but also the less likely is institutional change in response to Europeanisation. For example, Duina (1999) ascribes a critical role to the parliaments as determinant of institutional fit arguing they are 'guardians of the *status quo*'. Parliaments aggregate the opinions of all relevant stakeholders in one big forum and have an interest to maintain existing relationships. If EU policy requires a fundamental reorganisation of interest groups and their relationships to parliamentarians, policy is less likely to change.

Börzel & Risse (2013) argue that the goodness of fit depends on the compatibility with existing policies. They introduce two theoretical approaches, one rooted in 'the logic of consequentialism' and the other rooted in the 'logic of appropriateness. For both approaches they devise two mediating variables that determine whether misfit actually induces institutional change in response to a European law. While from the rational-choice perspective, the number of veto points and the formal institutions are the key factors that explain variation in implementation, from the sociological perspective, so-called change agents and the domestic political culture are key constraining factors.

The goodness-of-fit hypothesis was subject to criticism as it repeatedly produced null findings. Critics argued that the hypothesis is overly deterministic as it assumes a status quo bias of national governments (Mastenbroek & Kaeding, 2006). As an alternative Mastenbroek & Kaeding (2006) suggest to directly focus on the domestic preferences as policy change may come about as result of changes in domestic preferences that are influenced by a vector of inputs. Proponents of the hypothesis argue that, while misfit between European and national provisions is a necessary condition for change, it is not sufficient. The degree of change is then ultimately determined by domestic variables (Börzel & Risse, 2003; Knill & Lenschow, 1998). In a sense, Mastenbroek & Kaeding (2006) argue along similar lines as they emphasise the content of an EU proposal as a key determinant of change. However, they also show that misfit as a necessary condition does not generate explanations for cases of overcompliance. Sometimes compliant national governments use EU policy pressure as leverage for even more far-reaching ambitions. Due to the implicitly assumed status-quo bias of national governments, the pure goodness-of-fit model would fail to account for such cases.

Acknowledging these critiques, this thesis holds that, by cancelling out misfit as a necessary condition and instead directly focussing on the domestic variables, one can improve the explanatory value of Börzel & Risse's (2003) rational-choice-framework. While in cases of misfit, an EU proposal, in fact, exerts adaptational pressures, these modifications allow to also account for cases of overfitting after the third electricity directive. The following two sections present an integrated framework using the domestic variables of their framework and the notion of path dependencies as devised by Douglass North (1990).

Rational-Choice Institutionalism

"...the logic of rationalist institutionalism suggests that Europeanization leads to domestic change through a differential empowerment of actors resulting from a redistribution of resources at the domestic level (Börzel & Risse, 2003, p.2)."

Börzel & Risse's (2003) model starts from the presumption that Europeanisation must meet a certain 'inconvenience' or 'misfit' with existing power configurations to induce domestic change. While this thesis does not consider it a necessary condition, it acknowledges that an EU directive does exert pressures on domestic actors. Even in compliant cases, governments may use an EU policy proposal as a window of opportunity to introduce more extensive reform. Hence, reform, regardless where its main impetus comes from, would lead to a redistribution of power and a "differential empowerment of actors" (Börzel and Risse, 2003, p.2). As touched upon earlier, their rational-choice model identifies two mediating variables which determine to what extent this pressure really leads to domestic change: the number of internal veto points and the formal institutions present in a country.

17

Multiple Veto Points

Drawing on Tsebelis' (1995) veto-player theorem, the authors establish a link between the number of *de facto* veto players such as institutions or powerful interest groups and the difficulty to achieve the 'winning coalition' needed to introduce reform in response to European legislation. It is quite intuitive to assume that the more different actors are present in a system, the more dispersed are preferences across a diverse set of interests (Börzel & Risse, 2003). There may be situations in which the preferences of these actors converge on a given issue, dampening the constraining effect of multiple veto points. Nevertheless, the probability of preference convergence naturally decreases, the more actors are involved in the policy process (Tsebelis, 1995). Regarding the issue of ownership unbundling, it is not far-fetched to assume that multiple veto players create a barrier to institutional change. The electricity sector is subject to strategic interactions between various stakeholders at all levels of governance (Pelkmans & Lusetta, 2013). These barriers may be further accentuated by party-political differences between the involved political actors. (Tsebelis, 1995).

Formal institutions

Next to institutional and factual veto players, formal institutional arrangements may also provide actors with the ideational and material resources to exploit political opportunities. The authors argue that actors are more likely to exploit opportunities if existing formal rules or institutions are supportive of the intended action (Börzel & Risse, 2003). As such, certain laws can serve some actors as a catalyst for political opportunities while for others it acts as a constraint. In the complex regulatory environment of electricity regulation, intervention of the EU can quickly lead to conflict with existing national provisions and thereby create political clout for actors opposed to a harder unbundling regime.

In sum, from a rational choice perspective, adaptational pressures resulting from European reform intentions provide some actors of a domestic political system with opportunities while introducing constraints on others. The ability to circumvent these constraints or exploit the opportunities is highly contingent on the two mediating variables introduced above (Börzel & Risse, 2003). While this thesis considers these as the static constraints, the next section shows what I will call the dynamic or 'path-dependent' constraints.

Path Dependence

In this thesis, the notion of path dependence forms the dynamic constraints that affect the action scope of a given institutional actor. Path dependence plays an important role in the social sciences and is the central concept in historical institutionalism. In broad terms, the approach looks at the conditioning effects of an individual's past decisions on his or her available choice set in the present. While there is a vast body of conceptualisations of path dependence, the model used in this thesis mainly relies on the approach developed by Douglass North. Drawing on insights from technological development paths, North argues that actors become increasingly used to particular forms of organisation or problem solving if they have an experience of doing things in this way. Over time this may lead actors to favour this known path when confronted with the choice of multiple options even if the alternative generates more individual utility (North, 1990). This is often explained by increasing returns of the chosen path which makes the known option in the short-term *de facto* less costly as pre-established synergies result in lower short-run transaction costs (Pierson, 2000; Bernhard, 2014).

The first step triggering a path dependent process is known as *critical juncture*. The decision taken at this point in time sets in motion a chain of consequences involving a number of *feedback loops*, which over time consolidate the chosen path and make deviation gradually more unlikely. Eventually the individual or organisation almost irreversibly locked itself in the initial decision. Self-reinforcing mechanisms or feedback effects are critical aspects in this process. Feedback effects of a chosen path lead actors to hold on to this route. Examples of feedback can be learning effects or synergy effects (North, 1990; Bernhard, 2014).

The following section presents the variables and hypotheses that are derived from this theoretical framework.

III.   Research Design

Variables and Hypotheses: The Determinants of Unbundling Regime Choice

Taking the elaborations of the theoretical chapter as a starting point, this thesis argues that historical path dependencies and rational cost-benefit calculations may help to explain the degree of unbundling employed by a national government. More precisely, the ownership structures prior to first unbundling, regulatory schemes prior to regulation and internal veto players are thought to be important factors determining the compatibility of an unbundling regime with existing governance structures. Table 3 below illustrates the theoretical logic adopted in this paper and shows how this can be applied to the postulated variables. The following section presents the variables and hypotheses in more detail.

Ownership Structures

There are crucial differences between ownership unbundling of a private and a public VIU. As ownership unbundling may involve the effective expropriation of a private enterprise, it may present a serious clash between domestic law and European law and may contradict the fundamental rights laid down in national constitutions and the ECT, if international investors are involved (Hogan & Hartson Raue, 2007; Selivanova, 2010). In this case, private companies are likely to invest resources to prevent change and thereby push governments in long and costly rounds of negotiation. Moreover, even if a government successfully forces a private entity to give up the networks, the reparation to be paid may quickly overstretch the public budget. Several European (sub-)governments privatized shares of VIUs including the networks long before liberalisation. Following the historical institutionalist logic of path dependencies, this thesis argues that these states locked themselves in this decision and now have difficulties to carry on with hard unbundling. As opposed to Meletiou et al. (2018), I argue that the mere existence of private ownership and not majority private ownership determine governments' unbundling regime choice. This focuses more explicitly on potential legitimacy issues and costs of negotiation faced by a government if it instructs forced expropriation. As these are likely to occur even under minority private ownership, the following hypothesis is formulated:

H1: If VIUs were not fully state-owned prior to first unbundling, a government is more likely to adopt soft unbundling

Internal Veto Players

Internal veto players are one of the two mediating variables within Börzel & Risse's (2003) rational choice framework. The number of veto players and their strength varies across countries and political systems. In federal states, one may expect stronger constraints than in more centralised governments as sub-state entities are directly represented in the central government-making process. This often leads to the so-called *policy inertia* or policy gridlock federal systems are often accused of (Scharpf, 2011). In the present case, the motivation to block decisions is amplified if the sub-state actors have a stake in VIUs themselves. Therefore, the degree of redistribution of power in the present case is likely to be contingent on the existence of lower-tier government ownership with veto powers. In order to limit the loss of powers following liberalisation, the sub-states make use of their veto powers. As they are unlikely to give up their networks as a reliable source of income, the following hypothesis can be made:

H2: If sub-state actors with veto powers in the central policy process had ownership of a VIU prior to first unbundling, a state is less likely to adopt a 'hard' unbundling regime..

In addition, veto player theory predicts that convergence of preferences across legislative branches increases the possible win-set of the status quo and thus facilitates policy change (Tsebelis, 1995). In the present case, preferences may to a large extent be explained by material costs and benefits of all actors that have a stake in the utilities. However, it is not helpful for this analysis to completely ignore party-related preferences of political veto players. After all, a liberal party is still more likely to favour liberalised network sectors than a leftist party. On the other hand, the left party may favour the break-up of quasi-monopolies by unbundling their generation and network segments. Moreover, it could be imaginable that parties deviate from their real preferences for power-political purposes. As such, ideological congruence across legislative branches can have a catalysing effect for policy change. The following hypothesis is formulated:

H3: If the executive and the legislature were not dominated by the same party(-coalition) prior to first unbundling, a state is more likely to adopt a 'soft' unbundling regime.

Regulatory Schemes

Transmission systems are still considered natural monopolies due to economies of scale and long-term concessions. To prevent them from charging discriminatory network fees, states use various regulatory mechanisms to mimic competition (Joskow, 1996). Generally, two models can be described: cost-based and incentive-based schemes. Cost-based models use the marginal cost including a guaranteed rate of return on used capital as basis for the price. Incentive-based regulation, on the other hand, promotes efficiency by using price or revenue caps. Cost-based regulation is traditionally associated with a bias in capital investment due to the capital-premium, also known as Averch-Johnson effect. In vertically integrated utilities, revenue generated in production could be used for capital investment to increase network access fees to crowd out competitors (Meletiou et al., 2018). As such, transmission operators do not only have a more stable source of income but also more price-setting capacity under cost-based regimes. Moreover, hybrid schemes exist, where elements of both are used. One such example would be sliding-scale regimes, where prices are fixed *ex-ante* but can be increased after a certain threshold is reached. Nevertheless, the efficiency requirement still exists, which is why hybrid schemes can be thought to have similar effects like fully incentive-based schemes (Goetz, Heim & Schober, 2014). The following hypothesis is thus generated:

H4: If a state used a fully cost-based regulatory scheme prior to first unbundling, it is more likely to use soft unbundling

Table 3. Determinants of Unbundling Regime Choice (created by the author)

Causal mechanism	Cause	Mechanism		Outcome
Theory	**Domestic Constraints**	**Costs of Adaptation**		**Degree of Institutional Change**
	Privatisation of VIUs prior to liberalisation	Rent-seeking activities to maintain status quo	Expropriation requires high compensation	Soft Unbundling
	Shared ownership of VIUs with sub-national veto players	Reliable source of income for public budgets	High degree of persuasion necessary to achieve consensus	Soft Unbundling
	Partisan non-alignment between executive and legislature	More resources necessary to persuade legislature		Soft Unbundling
	Choice for cost-based regulation	Guaranteed profit for VIUs	Rent-seeking activities to maintain status quo	Soft Unbundling

Methodology

Quantitative Analysis

The hypotheses formulated in this thesis will be tested by means of an explanatory mixed-methods design. In a first step, a quantitative analysis will be conducted to check whether the theorised mechanisms are empirically observable across a wider range of countries. The sample includes 26 EU countries excluding Cyprus, Luxembourg, and Malta as these three countries have no high-voltage transmission lines and thus do not fit the parameters of this research. They are replaced by Albania, Norway and Switzerland which are no members of the EU but functionally associated with the internal market of electricity through bilateral or intergovernmental agreements and generally expected to adhere with the stipulations set out in the Third Electricity Directive[3].

Due to the low number of cases, the available options for elaborated statistical analysis are quite restricted. A Fisher's exact test satisfies the demands of this thesis. The procedure is suitable to test the independence of two different groups with dichotomous outcome variables and with low expected values. As such, it represents an alternative to a Chi-square test where, by conventional wisdom, expected cell counts should be higher than 5 (Sprent, 2011).

For the dependent variable *unbundling regime,* a dummy variable will be coded where 0 represents hard unbundling and 1 represents soft unbundling. Hard Unbundling includes OU and ISO as both are generally considered to be strict. Soft Unbundling equals ITO as well as hybrid systems where ITO is present as visible in Germany. The mixed regime of the United Kingdom will be counted to Soft Unbundling as the mix consists of ISO and OU. The data was collected from Meletiou et al (2018) and various Commission reports (European Commission, 2016, 2014, 2007, 2003, 2002 and 2001).

The independent variable *ownership structures* will be coded into a dummy where the presence of private ownership of VIUs one year prior to first unbundling[4] equals 0 and full state ownership equals 1. This type of coding represents a change to the approach employed by Meletiou et al (2018) who distinguish between majority ownership of private companies and majority public ownership. As argued in this thesis, their operationalisation only partially captures the reality as majority ownership only implies how internal decision-making of an organisation on various matters takes place. As ownership unbundling involves forced expropriation, also

---

[3] Albania needs to comply with the *acquis communautaire* as accession candidate; Regarding Norway, the EEA has incorporated the directive into its treaties (Bjørnebye, 2019); The relationship with Switzerland is more tensed but the latter is taking actions to comply with EU provisions and a bilateral agreement is currently negotiated. For more information on Switzerland consult Van Baal & Finger (2019).

[4] First unbundling refers to the first adoption of an unbundling regime that goes beyond mere 'legal unbundling'. Therefore, the point in time for some countries dates back to before the Third Directive as they already satisfied these conditions earlier.

minority ownership can be expected to induce legal conflicts where private owners want to hold on to their assets.

For the hypotheses regarding veto players, two different independent variables are coded. First, a variable for *sub-national ownership with veto powers* is adopted, where 1 equals ownership of provincial or municipal public entities. This is justified on the basis that not all sub-national entities have equal capacities to influence policymaking at the national levels. For example, provinces in centralised countries are thought to be overruled very easily by the central government while federal provinces often have the capacity to directly influence the central process. The data to check for sub-state ownership was collected from the VIUs annual reports and media articles (Verbund AG, 2011; Axpo Holding, 2014; Elia, 2012; Rosenberger, 2012).

Second, a variable for *partisan alignment* across veto players will be coded. This variable is inspired by the political constraints index by Henisz (2000) and seeks to explore the relationship between partisan alignment of policy-relevant veto players and capacity for institutional change in response to European legislation. For purposes of simplicity only two relevant veto players per political system are assumed: the executive and the legislature. In unicameral systems, a majority government will be considered as partisan alignment across legislative branches and thus coded as 1. In bicameral systems, partisan alignment is satisfied if the governing coalition has a majority in the lower house as well as in the upper house. If one of the two conditions is not satisfied, it will be assigned to a value of 0. This is an oversimplified view as it does not account for the degree of 'fractionalisation' of the legislature and views a multi-party executive as a coherent body without a significant degree of internal consensus finding. However, I believe that partisan alignment across legislative branches is more important for effective policymaking than internal coherence. Another limitation is the implicit assumption that the governing coalition is in favour of the most 'Europeanised' option, namely 'hard unbundling'. However, it might still fulfil the requirements of the hypothesis as in the event, it is in favour of this option, non-alignment can be an impediment to achieve that aim. The data for this variable is drawn together from the electionguide.org and in case of gaps the respective institutions directly.

The variable *regulatory scheme* is coded into a dummy where 0 represents an incentive-based scheme and 1 a cost-based scheme. Hybrid schemes are counted as incentive based as even the partial existence of this scheme shows it is more difficult for affected companies to be profitable. This is, again, a deviation from the operationalisation by Meletiou et al (2018) who distinguish three categories and include hybrid schemes.

Based on the significant variables of the analysis, I develop a decision-theoretic model to measure the interaction effects of the variables separately. The specifics of this model will be elaborated upon in the analysis section.

Comparative Case Study

To provide analytical depth to the findings of the quantitative analysis, a comparative case study of two similar cases will be conducted. This type of comparative study is useful to extract the variables that account for variation in outcomes between structurally similar cases (Goodrick, 2014). For the present analysis, case studies provide additional value by allowing to better contextualise the role of identified variables and establish their causality. Furthermore, qualitative analysis allows to demonstrate whether path dependent processes have also informed the unbundling decision of governments. A purely quantitative analysis can capture this aspect only to a limited extent. Another important benefit resulting from case studies is that they provide an additional opportunity to interpret the quantitative results. The analysis will concern the Netherlands and Germany which fundamentally differed in their unbundling choices but can be considered similar in many aspects. A more elaborate description of the case selection follows after the quantitative analysis.

IV. Quantitative Analysis

Table 4. Fisher's Exact Test of Unbundling Regime Choice (created by the author)

	Unbundling Regime				Odds Ratio	p
	Hard Unbundling		Soft Unbundling			
	N	**(%)**	**N**	**(%)**		
**Sub-national ownership with veto powers **					∞	0.017
yes	0	0.0	3	37.5		
no	20	100.0	5	62.5		
Partisan alignment					1.4	0.516
yes	14	70.0	5	62.5		
no	6	30.0	3	37.5		
Private ownership involved *					5.0	0.077
yes	5	25.0	5	62.5		
no	15	75.0	3	37.5		
Regulatory scheme **					9.44	0.022
Cost-based	3	15.0	5	62.5		
Incentive-based	17	85.0	3	37.5		
* Groups are significantly different $(p<0.1)$ ** Groups are significantly different $(p<0.05)$						

The first hypothesis *H1* was formulated the following way: *If VIUs were not fully state-owned prior to first unbundling, a government is more likely to adopt soft unbundling.* The Fisher's exact test above shows that the existence of private ownership is marginally significant with a p-value of 7%. Although only half of the observed cases which involved private ownership opted for a soft regime, the odds ratios show that in the presence of private ownership, the odds to maintain a soft unbundling regime are 5 times higher. For the present analysis, it may be reasonable to assume that private ownership did have an impact on the choice of several governments to opt for a softer unbundling regime. This interpretation will be more elaborated upon in the following parts of this thesis. Purely based on the statistical analysis, however, *H1* can only carefully be confirmed.

H2 posits that: *If sub-state actors with veto powers in the central policy process had ownership of a VIU prior to first unbundling, a state is less likely to adopt a 'hard' unbundling regime.* The Fisher's exact test proves this hypothesis with a p-value of 1.7%. The odds-ratio is infinity which can be explained by the non-existence of cases where veto player ownership occurred together with a hard-unbundling regime in the sample. This value can be misleading. Some authors suggest that this problem can be overcome by adding 0.5 to the empty cells (Ruxton & Neuhäuser, 2013). In this case the odds ratio would take the number of 24. While the exact value of these numbers should be regarded with caution, the odds ratio as well as the p-value imply a strong relationship between veto player ownership and a soft unbundling regime and thus *H2* can be accepted.

The third hypothesis *H3* related to the effect of partisan non-alignment on the decision to opt for a soft unbundling regime. A p-value of 51% suggests that there is no association between the two variables and thus the hypothesis cannot be accepted. Also, the odds ratio of 1.4 is rather small and suggests that both groups are independent. As such, these predictions of veto player theory do not hold in the present context.

The fourth hypothesis *H4* holds: *If a state used a purely cost-based regulatory scheme prior to first unbundling, it is more likely to use soft unbundling.* The analysis above shows strong evidence for this hypothesis. A p-value of 2.22 % suggests that the probability that there is an association between the two variables is high. 85% of all hard-unbundling cases included incentive regulation while almost 2/3 of all soft unbundling cases were in a cost-based regime. The odds ratio implies that the odds of states using a purely cost-based regime prior to regulation to opt for a soft unbundling regime are 9.44 times higher than for states using hybrid or purely incentive-based schemes. This suggests that price and revenue caps do indeed disincentivise states to maintain their networks and as such hypothesis H4 can be accepted.

It may be important to note that the chosen procedure does not account for the marginal effect of private ownership when controlling for other variables, nor for the interaction with other variables. For instance, does the presence of incentive regulation affect all countries similarly or does its effect depend on the ownership structures in the country of interest? To obtain a more nuanced understanding of the different drivers it may be necessary to account for all possible interaction effects. The next section introduces a simple theoretical model that attempts to fulfil this task.

Decision-Theoretic Model

To analyse the assumptions of Börzel & Risse's (2003) model in a more systematic way, this thesis proposes a formal decision-theoretic model. This removes ambiguities and allows to take interaction between variables beyond mere speculation. The model seeks to formally identify the utility-maximising strategies for a government's unbundling regime choice in any given configuration of the statistically significant variables. Several assumptions are made: First, although the model is concerned with sequential strategic interactions between several players, the situation is conceived as a single-player game, where the government responds to its pre-existing environment. Various scholars hold that sequential games can be viewed through single-player decision-analysis frameworks under the condition that the actions of the other players are reflected in the decision-makers payoffs (Van Binsbergen & Marx, 2006). Second, the governments' preferences are described as maximising aggregate social welfare to the largest possible extent. While in reality some governments may deviate from that, it can be considered a principal goal of any democratically elected government. Following these assumptions, under no constraints, a harder unbundling regime should be in the interest of any government as economic theory and scholarly evidence suggest that it benefits countries in the long-term. However, the cost-benefit calculation of the governments are short-term oriented and therefore are constrained by the transaction costs that are induced by the identified constraint variables. Third, the decision environment is considered deterministic and thus actors have complete certainty about the consequences of their choices. This means that a decision-maker can perfectly weigh the costs and benefits of any move she takes, and in any given situation, there is an objective single best move.

Based on these general assumptions, a decision-tree has been created where a government can be confronted with a total of eight different configurations of its environment (see Figure 3). Table 5 presents the various optimal choices in any given configuration. The results have then been made based on the intensity of constraints which are defined as: single-constraint, double constraint, and no constraints. Single constraints refer to the situation where either private actors or sub-state actors seek to prevent policy change. Double constraints refer to situations where both players invest resources in preventing change. In both cases the payoffs can be described as $B_i > A_i$ meaning soft unbundling is more short-term utility maximising. Only in 'no constraints'-scenarios, the government's optimal choice given the above made assumptions would be $A_i > B_i$ and thus favour soft unbundling. A main assumption behind the reasoning is that private actors only worry about monetary gains while sub-state veto players also care about power political gains. As such, price- or revenue-cap regulation would disincentivise private actors while veto players can still be expected to maintain VIUs even if they become less profitable as they would sacrifice a considerable degree of autonomy and bargaining power vis-à-vis the central government. While it is acknowledged that these assumptions are overly simplistic, they can be used as a starting point for more elaborate analyses.

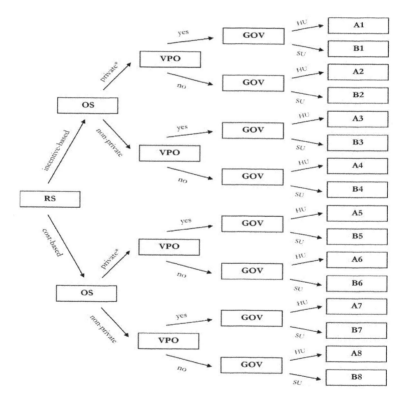

Figure 3. Decision-theoretic Model (created by the author) *(RS = Regulatory Schemes; OS = Ownership Structures; VPO= Veto Player Ownership; GOV=Government)*

Table 5 Sub-game results of decision-theoretic model (created by the author)

Sub-game	Best Decision	Reasoning
1	$B_1 > A_1$	**Single constraint:** sub-state actors insist on ownership despite revenue-caps (power-political purposes); Incentive regulation too threatening for private actors
2	$A_2 > B_2$	**No constraints:** Incentive regulation too threatening for private actors
3	$B_3 > A_3$	**Single constraint:** sub-state actors insist on ownership; private ownership not present
4	$A_4 > B_4$	**No constraints**
5	$B_5 > A_5$	**Double constraint:** guaranteed profit for both private actors and sub-state actors
6	$B_6 > A_6$	**Single constraint:** Guaranteed profit for private actors
7	$B_7 > A_7$	**Single constraint:** Guaranteed profit for sub-state actors
8	$A_8 > B_8$	**No constraints**

The model was tested in a simple way. Two groups were created where one group summarises all identified situations in which hard unbundling generates the highest payoff and one group for all configurations in which the model predicts that soft unbundling generates the highest payoff. By means of a Fisher's exact test, the model predictions were compared against the real outcomes. Table 6 shows that the model is significant at the 5% level what allows for the careful interpretation that it is able to capture the drivers of unbundling regime choice. However, as the upper left and lower right quadrants in Figure 4 show, there are several outliers which would require individual examination. A remarkable characteristic of three out of four countries where a hard regime was falsely predicted is that they used a cost-based network regime prior to first unbundling. Those same countries are also post-communist. Neither of these two features, however, can be plausibly used to explain their choice. Regulatory schemes are within the scope of the central government's competences and should not act as a constraint on the government. However, the emphasis on the communist history of some countries in their unbundling regime choice has also been done by Van Koten & Ortmann (2008). More elaborate models could control for the marginal effect of post-communism. Also, qualitative case studies may help to understand the outliers.

Qualitative analysis might also be useful to understand how the identified variables affected government's implementation of the Third Directive that were correctly predicted by the model. Can differences between structurally similar countries be plausibly explained by the factors that were determined in the quantitative analysis above? The following section addresses this puzzle by comparing the situations in the Netherlands which used the Third Directive to finalise distribution network ownership unbundling and Germany which still has a soft transmission unbundling regime.

Table 6 Fisher's exact test of decision-theoretic model (created by the author)

	Unbundling Regime				p
	Hard Unbundling		Soft Unbundling		
	N	(%)	*N*	(%)	
Model prediction **					0.038
Hard Unbundling	18	90.0	4	50.0	
Soft Unbundling	2	10.0	4	50.0	
* Groups are significantly different (p<0.1) ** Groups are significantly different (p<0.05)					

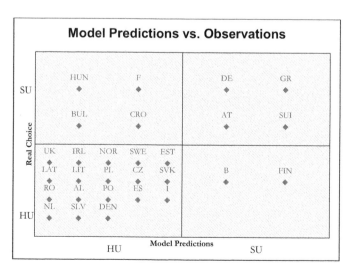

Figure 4 Predictions vs Observations decision-theoretic model (created by the author)

V. Qualitative Analysis

Case Selection

As the quantitative analysis can only provide us with correlational measures, it is now interesting to unpack how the identified variables impacted government decisions in two different cases. This has the benefit of providing a more nuanced understanding of the causality between the variables. More importantly, by means of the qualitative analysis, the role of path dependencies in governments' unbundling regime choice, especially with regards to ownership structures can be demonstrated. The process leading up to the unbundling regimes in Germany and the Netherlands will be analysed and compared. It is argued that these two cases represent a most-similar systems design. Both countries are similar in terms of economic power, have a similar culture due to geographic proximity and a shared history, have a multiparty parliamentary democracy and have low corruption perception indices (CPI). The last point was conceived by Van Koten & Ortmann (2008) to be a major explanation for variation in unbundling regime choice in Western European countries. It is important to note that the Netherlands already adopted ownership unbundling of TSOs after the first electricity directive. The third directive served the government to make advances in unbundling of distribution system operators as the first country in the European Union. As such, the analysis will focus on this particular outcome in the Netherlands and examines the factors that enabled the country to take unbundling so far. It should be noted that *H3* will not be explicitly tested as the quantitative analysis did not provide evidence for it.

Table 7. Case selection (created by the author)

	Germany	Netherlands
Features		
GDP/capita	A	A
Corruption	A	A
Culture	A	A
Euroscepticism/Compliance	A	A
Unbundling Regime	A	B

The analysis starts by analysing the case of Germany and then analyses the case of the Netherlands. The data is sourced from various sources but mostly relies on media articles, position papers as well as academic literature.

Case Study Germany

Ownership structures

From the early 20th century, the German VIUs were marked by shared ownership structures between the public – municipalities and the *Länder* - and the private sector. For the public sector, ownership was beneficial in various ways: on the one hand, it allowed them to have a stable source of income. In municipalities, for example, earnings from network industries were often used to finance investment in public transport or other obligations. On the other hand, ownership equipped them with political autonomy in a key sector. Due to this strategic importance, the liberalisation plans by the Commission were for a long time resisted by officials in Germany (Kleinwächter, 2012).

In the 1990s a significant privatisation wave could be observed in Germany. Two interrelated developments were key factors behind that and can be interpreted as critical junctures for the country's difficulties in unbundling the system. First, the systemic transformations in East-Germany following German re-unification led to a fundamental reorganisation of the electricity sector. The whole eastern economy was subject to privatisation processes by means of the *Treuhandanstalt* and the electricity sector was no exception to that. Especially investment in the ruinous transmission lines was difficult to accomplish without the financial means and know-how by private companies. As a consequence, a consortium of three giant electricity producers RWE, PreussenElektra and Bayernwerke AG gradually secured their influence in this new market (Birke, Hensel, Hirschfeld & Lenk, 2000). CEO of Bayernwerke AG, for example, highlighted the opportunities to further expand towards Eastern Europe: "If you have your thumb on the GDR network, you are also in good company with your targeted neighbours (Zeit, 1990)." As a consequence, the market concentration increased in Germany after reunification and this market concentration, in turn, made investment in RWE more and more attractive for investors.

The second factor was the diminishing profitability of the electricity business for some sub-state- and municipality-owned VIUs in the West and the East. Several cities and *Länder* sold their VIUs in the mid-1990s. A very prominent case was hereby the sale of BEWAG to RWE in Berlin in 1996. While the electricity system caused high costs on the local administration's budget, it increasingly came under public pressure to deliver on its election promises, in particular investment into the old transport infrastructure (Der Spiegel, 1996). Selling off the company to finance these promises thus served as short-term strategy to maintain legitimacy vis-à-vis the citizens. In addition, the administration feared that the European Commission's liberalisation plans, and the entry of new competitors could undermine the profitability of their business. The process was copied by several other utilities in various German cities and Länder with a similar situation (Der Spiegel, 1997). This sub-state overreaction is likely to have restrained the German central government's subsequent unbundling plans as this involved releasing the transmission lines into the hands of expanding privately-owned businesses and consequently increased their bargaining power.

The privatisation and internationalisation processes were further accelerated by the first liberalisation directive in 1996. As a result, the number of the previously eight large companies was reduced to the so-called 'four giants' RWE, E.on, Vattenfall and EnBW. The ownership of transmission lines allowed these giants to consolidate their market power by blocking access to the network and withhold investment for the expansion of new transmission lines. The obtained market power guaranteed investors' confidence in the business. In order to maintain this status quo, the incumbents used everything at their disposal to extract rents from the governments. The rules regarding 'third-party access' after the first directive demonstrate this well. As the only country in the EU, Germany opted for the model of negotiated access to the network instead of regulated access. Under this model, the network operators, producers and the energy-intensive industry negotiate access conditions multilaterally but without regulatory intervention. This allows the powerful utilities to negotiate favourable conditions for themselves. Fair 'third party' access is an important pre-condition for an effective unbundling regime, especially if there is still some degree of vertical integration present in the system (Brunekreeft & Meyer, 2009). The successful rent-seeking activities by the incumbents regarding general access rules exemplify their important influence in the policy process.

The second directive in 2003 of the EU required a more stringent regulatory regime where an independent regulator oversees network access and introduced *legal unbundling* as a minimum level of unbundling. Back then, the Commission already strongly advocated for the adoption of full ownership unbundling. However, this was met by vigorous resistance by the incumbent utilities. The director of the German federal cartel office Ulf Böge argued that a hard-unbundling regime will not be possible in Germany because of private network-ownership and underlined that forced expropriation is not an option for them (Energie und Management, 2002). This statement proves that the ownership structures are among the major factors preventing ownership unbundling in Germany.

Looking at the German response after the Third Directive, it becomes clear that their opinions were heavily influenced by RWE, Vattenfall, EnBW and E.on. It is fruitful to look at the process of policy formulation prior to the adoption of the directive to understand private companies' influence on the German decision adopt the ITO model which is in most aspects identical to legal unbundling. Germany led an alliance that heavily advocated for the ITO model next to the hard regimes proposed by the Commission (European Daily Electricity Markets, 2008). When comparing the German negotiation position with the argument presented by the big four utilities, it becomes clear that the latter has the capacity to shape the opinion of the former. The main arguments raised were that ownership unbundling will not bring about benefits in competition and end-user prices (EU Energy, 2006; EU Energy, 2007). In order to push through their demands, the companies employed heavy lobbying and rent-seeking activities, not only at the national level but also directly at the Commission. It is plausible to assume that the German government was not able to enforce stricter unbundling rules after the third directive as they already obtained several concessions by the big utilities. First, the creation of the *Bundesnetzagentur* and tightened third-party access rules after the second directive already made it more difficult for them to exploit their market power. Second, incentive regulation was introduced in the same year as the third directive (Der Spiegel, 2007).

The analysis of ownership structures of VIUs in Germany, in fact, prevented the adoption of a stricter unbundling regime. The privatisation wave in the 1990s that was accelerated by the German reunification process can be considered a critical juncture that triggered a path-dependent process that almost irreversibly limited the German government's options to respond to European electricity regulation. When combined with the findings from the Dutch case, the analysis demonstrates the causal mechanism that explains the relationship between the two variables. The next part seeks to uncover the mechanisms for the other two variables that were identified in this paper.

Regulatory Schemes

As elaborated earlier, network access and its price were negotiated between the affected parties. However, the politically determined prices did not bring about the efficiency benefits as hoped by the government and therefore electricity prices ranked among the highest in Europe in the early 2000s. Following the EU's 'acceleration directive' in 2003, the task of determining prices was delegated to the newly set up regulatory agency *Bundesnetzagentur*. The agency introduced a cost-based scheme where the network price is accounted for by the expected capital costs, depreciation and a guaranteed rate-of-return. However, the scheme was not successful in reducing the costs for the end consumer. In practice, the regulatory agency struggled to prove inefficiencies on the side of the large network operators which allowed the latter to keep overinvesting and crowd out competitors (Kleinwächter, 2012).

In 2009, the government thus introduced incentive regulation. While this was met by the resistance of most network operators, for the present analysis it makes more sense to consider the effects of cost-based regulation as this was the scheme that was active when the Directive was decided and thus affected the German government's choice. Strikingly, in 2010, one year after the introduction of incentive regulation, E.on and RWE gave up their transmission network although they heavily advocated for the soft ITO-model (Biermann, 2009). Although EnBW is still vertically integrated down to the present day, this decision shows that incentive regulation, in fact, can make the network business less profitable. One may hypothesise that if the German government had been able to introduce incentive regulation earlier, the third electricity directive may have already served them as an opportunity to introduce a hard-unbundling regime. More interestingly, E.on's decisions demonstrates that the German government's decision was, in fact, heavily constrained by the companies and that its regulatory capacity is subject to their interests.

Veto Players

The German political system is marked by a type of 'cooperative federalism'. The *Länder* have a right to initiative by means of the *Bundesrat*, they can impact central policy making through various cooperation bodies and they also have a high degree of autonomy in policymaking for themselves (Monstadt & Schneider, 2016). As such, any initiative by the EU will have to overcome many instances before it can be translated into national policy. Regarding the issue of ownership unbundling and electricity liberalisation in general, federalism constituted a real impediment to progress. This applies, in particular, to the asset preservation ambitions by sub-state actors. For example, the state of Baden-Württemberg, which is home to a large part of the energy-intensive industry, owns the majority of shares of EnBW including its associated transmission network operator TransNetBW. The state directly owns 47 % of the shares while another 47% are in the hands of local authorities. Both parties agreed to take mutual votes in important decisions, implying a large degree of coherence. The strategic importance of electricity policy was articulated by regional politicians in various public communications (Rosenberger, 2012; Freytag, 2010).

Although in the other states only indirect ownership[5] of VIUs was present, they still had an interest in blocking more far-reaching unbundling efforts. For example prior to the third directive, the delegation of the state of Northrhine-Westphalia lobbied the Commission to tell them that ownership unbundling does not entail benefits in terms of market integration and that it only leads to unnecessary expropriation (WWU Münster, 2007). Accordingly, the Bundesrat's negative attitude towards the Commission's unbundling plans do not come as a surprise. Besides stressing that the status quo is sufficient for achieving the competition aims, the chamber argued that it is not clear who is to pay for compensation of private stakeholders in the event of full ownership unbundling. Moreover, they voiced their discontent that EU legislation gradually erodes their own legislative powers (Bundesrat, 2007). This is even more so the case, if they have to surrender assets under their own ownership.

---

[5] Ownership by local authorities

Another argument raised in the Bundesrat communication was the fact that incentive regulation was to be introduced and that one should first try to see whether this brings about the desired changes. Their communication implies that incentive regulation had already been a concession and that they are not willing to give in another time. Baden-Württemberg and Northrhine-Westphalia have been outspoken critics regarding incentive regulation but were not able to forge a majority on the issue (Der Spiegel, 2007). As it appears, however, regarding ownership unbundling the Bundesrat acted as a coherent body with a very clear stance.

Shortly after the directive was implemented as part of a comprehensive amendment of the *EnWG* in 2011, RWE and E.on introduced the ownership unbundling model by respectively selling the transmission system to Dutch company TenneT and setting up transmission operator 50 Hertz (Biermann, 2009) One year later Vattenfall implemented the new provisions (Frankfurter Allgemeine Zeitung, 2010). While Vattenfall also applied hard unbundling, only EnBw maintained the softer model (EnBW, nd). While for the other companies, incentive regulation may have given the final impetus to voluntarily give up the networks, the state of Baden-Württemberg seemed to want to maintain its strategic edge resulting from VIU ownership.

From the elaborations above, it is evident that next to private interests and the presence of cost-based regulation, the German *Länder* acted as a strong constraint on the implementation of the third electricity directive. The case study is in line with the expectations for Germany from the quantitative analysis. To boost the causal inferences to be drawn from that case study, the following chapter analyses the case of the Netherlands which differs in the three identified factors as well as in the outcome variable.

Case Study Netherlands

Learned from past mistakes? TSO-Unbundling in the Netherlands

As the Netherlands had already unbundled their transmission network in the early 2000s, the third electricity directive served the government to finalise ownership unbundling of their low-voltage distribution networks. As such, the case study focuses on the interaction of the identified variables with the process of distribution network unbundling in the Netherlands. Nevertheless, it is useful to first outline the process of transmission system unbundling as this is inextricably linked to the subsequent separation of distribution systems. The following section shows how early liberalisation experiments may have informed the extraordinarily progressive development of the Dutch electricity system with regards to unbundling.

Contrary to most other European countries, the Netherlands articulated first ambitions to liberalise their electricity systems already in 1989, when the electricity act was formulated. The act formulated two clear goals for the Dutch electricity network: First, it sought to increase competition in the generation sector. Second, it has tried to make generation more efficient by coordinating production between the large-scale producers. The sector was at this time controlled by big four regional operators, each of which under the control of the regions and provinces: EPZ, EPON, ENA and EZH (Van Damme, 2005). The new law anticipated more competition by allowing distribution operators and industrial consumers to buy electricity from different producers instead of the local generating company. In addition, it made decentralised production capacity available and thereby allowed distribution companies and the energy-intensive industry to produce electricity themselves. This was promoted by forcing distribution companies to buy every unit of electricity at a fixed price (Van Damme, 2005).

In order to make generation more efficient, the electricity act established the SEP (Samenwerkende Electriciteits Produktiebedrijven[6]) which consisted of the four big electricity producers. The SEP had full ownership over the transmission lines and cross-border interconnections. By means of coordination, it was hoped to keep dispatch costs at the lowest possible level. The SEP had various important roles in the electricity system. From now on capacity planning had to be approved by the SEP as well as the minister. As SEP controlled the transmission lines, it acted as a clearing board between generation and supply. It can be argued that the electricity act already induced a certain degree of disintegration (Brunekreeft, 1997).

However, the electricity act appeared to be fraught with several design mistakes and inconsistencies. The granted feed-in tariffs for decentralised capacity soon induced an overinvestment into decentralised production. Due to the profitability, distribution firms increasingly merged with industrial producers and became multi-product firms. This led to a concentration of distribution companies and shifted market power towards this segment. This situation led to a under-utilisation of existing capacity by the large-scale producers, thereby increasing their cost per unit, which not only increased end-user tariffs but also increased feed-in-tariffs for decentralised producers as these were coupled to the average-price level of the big four. As a consequence, the market soon was marked by excess capacity in the system and significant price increases. Also, competition between the four large producers was not really incentivised by the new law as price-setting mechanisms essentially created a uniform tariff level. Put together, the new system appeared to be self-destructive (Brunekreeft, 1997; OECD, 1998).

The 'purple coalition'-government soon recognised these problems and anticipated them in the *Derde Energienota* (Third White Paper on the Electricity Sector) in 1996, paving the way for a new electricity law. The law had three goals: First, more rigorous measures to induce competition in the electricity sector. Second, creating the conditions for an open European

---

[6] English translation: Cooperating Electricity Production Companies

internal market of electricity. While the upcoming EC directive clearly guided the reform ambitions by the Dutch, the communication also noted that the wish for competition is accompanied by a fear of new market entrants from neighbouring European countries. Third, somewhat counterintuitively, a new law should address the issues of excess capacity by increasing vertical cooperation (Brunekreeft, 1997). The ambitions articulated in the White Paper show the dissatisfaction of the Dutch government with their first liberalisation experiment. It might be for that reason that the new law following the first electricity directive included more risk-averse measures and led to a harder form of transmission network unbundling, albeit not full ownership unbundling.

Following the new electricity law, the Dutch government set up TenneT as a new transmission operator and bought 50.1 % of existing transmission capacity. The rest remained in the ownership of SEP. While the new TSO had been structurally unbundled from the rest of the system, the ownership of lower-tier government through SEP remained intact. As such, it does not come as a surprise that they continued to misuse their market power and discriminate access to the network. In 1999, a large industrial consumer made a successful legal claim against SEP at the cartel office, which argued that the ownership structures, in fact, are not conducive to competition. Following the judgement, in November 2001, the Dutch government bought the rest of the shares from TenneT and thereby effectively paved the way for transmission system ownership unbundling (Möllinger, 2009). This early decision illustrates the provinces' relative weakness in the Dutch political system. Having outlined the early Dutch transmission network unbundling, the following sections now turn to examining how the identified variables can be held as explanatory factors for the separation of distribution system operators (DSOs)

Ownership Structures and Veto Players and DSO-unbundling

The DSO-unbundling process in the Netherlands illustrates the relatively low bargaining power of Dutch provinces vis-à-vis the central government. In 2006, three years prior to the Third Electricity Directive, the Dutch government introduced the WON-act and thereby already tried to force ownership unbundling upon distribution network operators. This 'gold-plating'[7] was not welcomed by the four big companies, Essent, Eneco, Nuon and Delta which were still integrated in the production and distribution segment and fully owned by the provinces and municipalities(International Law Office, 2017). During the process leading up to the WON-act, it became publicly known that the utilities bribed the consultancy IMSA to lobby the Dutch government against ownership unbundling (Van Koten & Ortmann, 2008). This 'corruption-

---

[7] Gold plating refers in this context to Member States' practice to over-comply with EU legislation

scandal' does not only illustrate the value the utilities attach to integrated networks but also the weakness of the provinces vis-à-vis the central government. Following the Third Directive, another amendment of the Dutch electricity act then set the 1 January 2011 as deadline to comply with the WON provisions. While Nuon and Essent split their holdings by respectively selling their production facilities to RWE and Vattenfall shortly after the Third Directive in 2009, Eneco and Delta appealed against the unbundling decision in front of the Hague Court of first instance (Looijestijn-Clearie, 2014).

The provinces argued that the law infringed Article 49 of the TFEU relating to the freedom of establishment and against the free movement of capital as stipulated in Article 63 TFEU. Relying on a provision that prohibits the privatisation of existing public assets, the Court of first instance ruled in favour of the Dutch government. However, when the companies brought the case to the Court of Appeal, the latter judged that the law is, in fact, an infringement of the free movement of capital. The Dutch government then appealed this decision in front of the Supreme Court, which referred the case to the CJEU for a preliminary ruling. In 2015, the CJEU ruled that the Dutch legislation is consistent with EU provisions (Looijestijn-Clearie, 2014; International Law Office, 2017)

The Dutch DSO-unbundling process confirms the expectation that sub-state actors in the Netherlands appear to be much less powerful than in federal states and can be overruled quite easily. Although the specifics of the eventual rulings are left to the discretion of the judges, the very fact that the cases were brought to court by the provinces is exemplary of their relative weakness vis-à-vis the central government. In federal states, sub-state actors can prevent laws from being introduced in the first place. Given the preferences of the provinces and the municipalities, it is not far-fetched to assume that the unbundling requirements would not have gone as far. While there is a certain formal degree of regional representation in the *Eerste Kamer*, various observations suggest that there is more alignment with the second chamber than with the regions (Source). Beyond these general power dynamics, regulatory conditions may have also influenced the Dutch success in DSO-ownership unbundling. The next part seeks to elaborate on this aspect.

Incentive regulation DSOs

The last section has shown that the central government of the Netherlands was able to impose its will upon the provinces. However, while Delta and Eneco engaged in a legal dispute, the restructuring of Nuon and Essent proceeded rather smoothly. The following section seeks to demonstrate how network pricing regulation may have contributed to the low resistance by the latter two companies. The Dutch system of incentive regulation is a model of yardstick competition. Network operators are incentivised to operate more efficiently by being subject to a revenue cap that is predetermined by the regulator DtE for a regulatory period of five years (prior to 2016, it was three years). The cap takes the average total cost of all DSOs in the year prior to the regulatory period as a baseline. Over the period, the allowed revenue decreases annually by a so-called x-factor that accounts for efficiency gains that network operators are expected to realise through technological advancement and smart investment decisions. The x-factor also adjusts for inflation. In addition, a quality-factor is included in the tariff calculation methodology and remunerates network operators for improvements in network quality, where quality is accounted for by the annual minutes of power outages in relation to the country-average (SAIDI-index). As such, better than average network quality means more revenue, worse than average network quality equals a lower revenue cap (Autoriteit Consument & Markt, 2017; Hesseling & Sari, 2006).

While incentive regulation is useful to keep lower end-user tariffs, it may make business less profitable for network operators and lead to the eventual sale of networks. Therefore, it is not far-fetched to assume that the Dutch advances in DSO-ownership unbundling is also partially related to the existence of its quite strict yardstick competition regulatory regime. As Figure 5 illustrates, available funds of DSOs substantially decreased after the introduction of incentive regulation. While investment levels did not suffer so much under the new regulatory scheme, operators reported that they only made "necessary investments" (Haffner, Helmer & von Til, 2010). This sentiment shows that the operators increasingly suffered under the regulatory constraints imposed upon them. According to Haffner et al (2010), 50 per cent of the operators said that they are afraid that in the future, the allowed revenue will not be enough to cover their costs. The price they pay for the regulatory burden can also be seen by the deteriorating quality of the distribution system. As can be seen from Figure 6, the SAIDI increased constantly in the period between 2001 and 2010. While not exclusively attributable to incentive regulation, it reflects additional pressure on the network operators and thus implies a lower incentive to maintain networks.

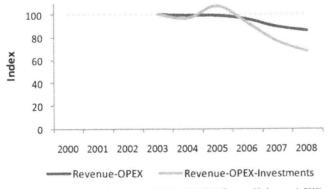

Figure 5 Revenue development DSOs 2000-2008 (Source: Hafner et al, 2010)

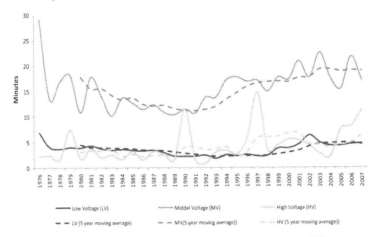

Figure 6 SAIDI Index-Netherlands 2000-2007 (Source: Hafner et al, 2010)

The increased pricing pressure soon led to plans for a merger of Nuon and Essent in 2007 (Handelsblatt, 2007). Two years later both, Nuon and Essent were sold to Vattenfall and RWE. As communicated by the provincial shareholders, the competitive pressures induced by foreign companies has become too strong (Aan de Brugh, 2009). While "it is difficult to disentangle the effects of privatization, restructuring and incentive regulation (Joskow, 2006, p.73)", the data shows that incentive regulation did exert significant price pressure on the incumbents, may have led to the fusion and eventually led to the sales of the production and supply companies. The distribution networks remained in the hands of the municipalities and provinces which effectively led to ownership unbundling. As the drivers of such strategic decisions are often difficult to disentangle, this qualitative analysis only allows for careful inference. However, the eventual outcome is in line with the quantitative findings and therefore underscores the hypothesis that revenue caps incentivise ownership unbundling of distribution networks.

Conclusion

Despite increasing convergence in the last three decades, internal interest and governance structures continue to restrict national governments' capacity to fully adapt to supranational reform pressures in the electricity sector. This thesis demonstrated this mechanism by answering the question: What explains the variation in European states' unbundling regime choice after the third electricity directive? Looking at unbundling regime choice to illustrate national electricity sector's transformative capacities was particularly intriguing for two reasons: First, due to market participants' dependency on the monopolistic transmission lines, vertical integration may lead to suboptimal market outcomes. Thus, the degree of unbundling can be considered a key indicator of a country's level of electricity market liberalisation. Second, forced separation of network ownership results in an unambiguously comprehensible short-term redistribution of resources between the parties affected by regulation. This allows us to view liberalisation and Europeanisation processes in light of governments' potential costs of adaptation, resulting from the compensation of losers of reform and the transaction costs involved in convincing them of the intended action in the first place.

The statistical results showed that the presence of VIU-ownership by sub-state veto players and private shareholders as well as the incentive structures provided by cost-based network regulation increase a member state's probability to maintain a softer unbundling regime. However, contrary to the expectations of veto player theory, party-political convergence across legislative branches did not have a significant influence. These findings were cross-validated and extended by a comparative case study between Germany and the Netherlands. It could be seen that governments' past decisions have a large impact on their ability to adopt a 'hard' unbundling regime. In the case of Germany, the government fell victim of a privatisation wave in the early 1990s, which strongly constraint its capacity to take more far-reaching measures. The Dutch case of overcompliance, on the other hand, showed how an anti-privatisation clause of public assets greatly facilitated the liberalisation process. Moreover, it demonstrated how the government may have been motivated to not repeat the mistakes of a failed liberalisation experiment prior to the EU directive.

The findings have various implications for research on EU member states' responses to the EU electricity agenda as well as for the temporal and spatial diffusion of liberalisation and Europeanisation more generally. They demonstrate how the implementation of an EU directive highly depends on the veto actors present in a policy community. If relevant players are affected by policy, the system is more resistant to change. This effect can be amplified if other legal obligations provide these actors with additional incentive structures. The interaction between regulatory schemes and VIU-ownership greatly demonstrate this connection and thus it appears that the domestic variables in Börzel and Risse's (2003) goodness-of-fit framework are able to generate accurate explanations for member states' variation in implementation of electricity legislation.

The findings are also consistent with Bartle's (2002) argument regarding the pace and timing of electricity sector liberalisation. In fact, the structural configuration of domestic institutions is a major determinants of national electricity system's short-term reform capacity. The thesis refined this argument by shifting the focus more explicitly towards the costs of adaptation to reform. Due to the vested interests of sub-state and private owners, in many cases, 'soft-unbundling' appeared to be the short-term equilibrium outcome. Considering the short-term oriented nature of decision-making, the argument of path dependent 'lock-ins' gains additional relevance. The compensation claims articulated by the private sector disincentivised governments to take the junction towards a fully liberalised. This relationship was visible in the case study of Germany. If the privatisation of VIUs had not occurred before liberalisation became a mainstream idea, the government may have had less problems to implement a 'hard unbundling' regime.

Furthermore, the findings highlighted the constraining role of federalism in national transformation processes. The entangled decision-making structures of federalism are often described by scholars to induce *policy inertia* (Scharpf, 2011). This in-built rigidity was also evident in the implementation of the third electricity directive. In Germany, due to the interests of the *Länder*, it was impossible to achieve majorities in the Bundesrat for ownership unbundling. The statistical results imply that this effect also holds in other countries where federal sub-states held ownership in VIUs. In contrast, the Dutch case showed that if lower-tier actors are not equipped with blocking mechanisms, the central government has much less efforts to impose its will upon them.

Next to its contributions to the field of Europeanisation and the institutionalist literature on network liberalisation, this thesis enriched our knowledge in the interdisciplinary research on unbundling regime choice. The introduction of sub-state ownership as a determinant of unbundling regime choice constitutes a novelty and this thesis suggests that future models benefit from the inclusion of this variable. Moreover, decision- or game-theoretic formalisation refines our understanding of the strategic interactions during the process of unbundling. The approach presented in this thesis can be a starting point for more sophisticated mathematical models.

Due to time, space as well as language constraints, the master thesis was not able to consider all relevant aspects of government's unbundling regime choice. For example, an empirical test of the role of national regulatory agencies' preferences as an additional veto player as suggested by Lindemann's (2015) model could have been interesting. However, due to language barriers, it would have been difficult to determine the preferences of every agency with certainty. On the other hand, the role of regulatory agencies is not to be overestimated as they have no direct ownership preferences as opposed to other actors. Nevertheless, empirical research could include this variable to further refine models. Another avenue for future research would be the testing of empirical models in other contexts outside of Europe.

Beyond their academic implications, the results of this thesis may also inform practical decision-making. The tensed role of sub-state and private ownership implies that central government ownership of transmission lines is likely to be conducive to making the process of unbundling as smoothly as possible. Governments outside of Europe that plan to liberalise their electricity sectors are therefore well advised to keep their grip on the networks.

References

Aan de Brugh, M. (2009, February 25). And everyone is suddenly fine with this takeover [En deze overname vindt iedereen opeens wel prima]. *NRC Handelsblad,*. Retrieved from https://www.nrc.nl/nieuws/2009/02/24/verkoop-energiesector-lijkt-niet-te-keren-11688434-a834158

Alesina, A., Ardagna, S., Nicoletti, G., & Schiantarelli, F. (2005). Regulation and Investment. *Journal of the European Economic Association*, *3*(4), 791–825. https://doi.org/10.1162/1542476054430834

Axpo Holding AG (2013). Annual- and Sustainability Report 2012/2013 [Geschäfts- und Nachhaltigkeitsbericht 2012/13], Retrieved from https://www.axpo.com/ch/de/ueber-uns/investor-relations/berichte-und-termine.html

Bartle, I. (2002). When Institutions No Longer Matter: Reform of Telecommunications and Electricity in Germany, France and Britain. *Journal of Public Policy*, *22*(1), 1-27. https://doi.org/10.1017/S0143814X02001010

Bernhard, P. (2014). Energy Policy in the Baltic States - Implementation of the EU's Third Liberalisation Package for the Natural Gas Market. [Energiepolitik imBaltikum - Umsetzung des 3. Eu-Liberalisierungspakets zum Gasbinnenmarkt in Litauen und Lettland]. (Diploma Dissertation). https://doi.org/10.13140/2.1.4851.5202

Biermann, K. (2009, February 28). E.on, these traitors [E.on, diese Verräter]. *Zeit Online*. Retrieved from https://www.zeit.de/online/2008/10/eon-politik-reaktionen/komplettansicht

Birke, A., Hensel, V., Hirschfeld, O. & Lenk, T. (2000). The East German electricity industry between public property and competition [Die ostdeutsche Elektrizitätswirtschaft zwischen Volkseigentum und Wettbewerb]. Leipzig. Retrieved from http://www.econstor.eu/bitstream/10419/52380/1/672206358.pdf

Bjørneby, H. (2019). The impact of the third energy market package on national resource management. Retrieved from https://www.energinorge.no/contentassets/bcee1a1d4bd842aaa7bf1b74f0441a07/legal-analysis-third-energy-market-package-080119.pdf

Bolle, F., & Breitmoser, Y. (2006) *On the Allocative Efficiency of Ownership Unbundling*, European University Viadrina Frankfurt (Oder) Department of Business Administration and Economics Discussion Paper No. 255.

Börzel, T. A., & Risse, T. (2003). Conceptualizing the Domestic Impact of Europe. In K. Featherstone & C. M. Radaelli (Eds.), *The Politics of Europeanization* (pp. 57-80). Oxford University Press. https://doi.org/10.1093/0199252092.003.0003

Brunekreeft, G. (1997). The 1996 reform of the electricity supply industry in The Netherlands. *Utilities Policy, 6*(2), 117–126. https://doi.org/10.1016/S0957-1787(97)00005-2

Brunekreeft, G. (2008). *Ownership Unbundling in Electricity Markets -A Social Cost Benefit Analysis of the German TSO* (EPRG Discussion Paper 08-15), Retrieved from https://www.researchgate.net/profile/Gert_Brunekreeft/publication/4999027_Ownersh ip_Unbuilding_in_Electricity_Markets_- _A_Social_Cost_Benefit_Analysis_of_the_German_TSO'S/links/572e355208aeb1c73d1 294ce/Ownership-Unbuilding-in-Electricity-Markets-A-Social-Cost-Benefit-Analysis-of- the-German-TSOS.pdf

Brunekreeft G., & Meyer, R. (2009). Unbundling on the European electricity markets: state of the debate. [Entflechtung auf den europäischen Strommärkten: Stand der Debatte]. In Knieps, G., & Weiß, H.-J. (Eds.) *Network economics case studies [Fallstudien zur Netzökonomie]*. Gabler.

Bundesnetzagentur. (2013) Unbundling models in the EU and certification of TSOs in Germany. Retrieved from https://www.ceer.eu/documents/104400/-/-/40770749-0a94- c65a- 1b02-5c7a9ec3aa58

Bundesrat. (2007). Drucksache 673/1/07. Retrieved from http://dipbt.bundestag.de/dip21/brd/2007/0673-1-07.pdf

Copenhagen Economics (2005), *Market Opening in Network Industries:* Part I Final Report, Copenhagen Economics for DG Internal Market, Retrieved from https://www.copenhageneconomics.com/dyn/resources/Publication/publicationPDF/1 /111/0/Market_opening.pdf

Cremer, H., Cremer, J. and De Donder, P. (2006), *Legal Vs Ownership Unbundling in Network Industries,* CEPR Working Paper, No.5767.

Der Spiegel. (1996, January 1). "Suicide out of fear" [„Selbstmord aus Angst"]. Retrieved from https://www.spiegel.de/spiegel/print/d-8871446.html

Der Spiegel. (1997, May 12). Negative Rest [Negativer Rest]. Retrieved from https://www.spiegel.de/spiegel/print/d-8715883.html

Der Spiegel. (2003, November 11). E.on increases network charges due to wind power / higher electricity prices likely next year. [E.on erhöht Netzentgelte wegen Windkraft/Höhere Strompreise im nächsten Jahr wahrscheinlich]. Retrieved from https://www.spiegel.de/spiegel/vorab/a-275197.html

Der Spiegel. (2007, September 21). States against lower electricity and gas prices [Länder gegen niedrigere Strom- und Gaspreise]. Retrieved from https://www.spiegel.de/politik/deutschland/bundesrat-laender-gegen-niedrigere-strom-und-gaspreise-a-507013-amp.html

Diekmann, J., Leprich, U. & Ziesing, H. (2007): Regulierung der Stromnetze in Deutschland. Ökonomische Anreize für Effizienz und Qualität einer zukunftsfähigen Netzinfrastruktur [Regulation of electricity grids in Germany. Economic incentives for efficiency and quality of a future-proof network infrastructure]. Düsseldorf: Hans-Böckler-Stiftung

Duina, F. G. (1999). *Harmonizing Europe: Nation-states within the Common Market. SUNY series in global politics.* State University of New York Press

Dutton, J. (2014). *EU Energy Policy and the Third Package.* University of Exeter Energy Policy Group Working Paper: 1505, Retrieved from https://ukerc.rl.ac.uk/UCAT/PUBLICATIONS/EU_energy_policy_and_the_third_package.pdf

Elia Group (2012). Annual Report 2011, Retrieved from https://issuu.com/elianv/docs/annual-report-2011

EnBW. (nd). *Geschichte der EnbW [History of EnbW].* Retrieved from https://www.enbw.com/unternehmen/konzern/ueber-uns/geschichte-der-enbw/

EnwG. (1998). Law for the new regulation of the energy industry law [Gesetz zur Neuregelung des Energiewirtschaftsrechts] (Energiewirtschaftsgesetz (EnWG), BGBl. II, S. 730-736

Ernst and Young (2006). *Final Report Research Project: The Case for Liberalisation*, Retrieved from http://www.dti.gov.uk/files/file28401.pdf

Energie und Management (2002). *Kartellamt mahnt ab [Cartel Office warns].*, Retrieved from https://www.energie- und-management.de/nachrichten/detail/kartellamt-mahnt-ab-40165

European Commission (2001). *Commission Staff Working Paper. First benchmarking report on the implementation of the internal electricity and gas market.* Brussels: author

European Commission (2002). *Commission Staff Working Paper. Second benchmarking report on the implementation of the internal electricity and gas market.* Brussels: author

European Commission (2003). *Commission Staff Working Paper. Third benchmarking report on the implementation of the internal electricity and gas market.* Brussels: author

European Commission (2007). *DG Competition Report on Energy Sector Inquiry 10.* Brussels: author

European Commission (2014). *Communication from the Commission to the European Parliament, the Council, the European Economic and Social Committee and the Committee of the Regions: Youth Opportunities Initiative.* Brussels: author

European Commission (2016). *Opinions and Decisions on Operator Certifications.* Brussels: author

European Daily Electricity Markets (2008, January 31). *France and Germany say ownership unbundling breaks EU law.* Retrieved from https://advance-lexis-com.ezproxy.ub.unimaas.nl/api/document?collection=news&id=urn:contentItem:4RRD-CYC0-TX4W-T1MV-00000-00&context=1516831.

EU Energy (2007, June 15). E.ON, utilities propose 'regional TSO' alternative to unbundling. Retrieved from https://advance-lexis-com.ezproxy.ub.unimaas.nl/api/document?collection=news&id=urn:contentItem:4P30-FHK0-TWMG-G2C3-00000-00&context=1516831.

EU Energy (2006, September 22). *Effective regulation preferable to ownership unbundling: E.ON.* Retrieved from https://advance-lexis-com.ezproxy.ub.unimaas.nl/api/document?collection=news&id=urn:contentItem:4M28-0P90-TWMG-G20W-00000-00&context=1516831.

European Parliament & Council of the European Union. (2009, July 13). Directive 2009/72/EC concerning common rules for the internal market in electricity and repealing Directive 2003/54/EC. *Official Journal of the European Union*, L 211/55

Frankfurter Allgemeine Zeitung. (2010, March 12). Vattenfall sells German high-voltage network [Vattenfall verkauft deutsches Hochspannungsnetz]. Retrieved from https://www.faz.net/aktuell/wirtschaft/wirtschaftspolitik/energieversorgung-vattenfall-verkauft-deutsches-hochspannungsnetz-1957208.html

Freytag, B. (2010, December 7). Modern industrial policy „Made im Ländle" [Moderne Industriepolitik „Made im Ländle"]. *Frankfurter Allgemeine Zeitung*. Retrieved from https://www.faz.net/aktuell/wirtschaft/unternehmen/mappus-bei-enbw-moderne-industriepolitik-made-im-laendle-1578910.html

Goetz, G., Heim, S., & Schober, D. (2014). Ökonomische Aspekte von Stromleitungsnetzen Ökonomische Aspekte von Stromleitungsnetzen [Economic Aspects of Economic aspects of power line networks]. In J. Böttcher (Ed.), *Stromleitungsnetze: Rechtliche und wirtschaftliche Aspekte* (1st ed., pp. 285–310). De Gruyter Oldenbourg.

Goodrick, D. (2014). Comparative Case Studies, *Methodological Briefs - Impact Evaluation No. 9,* Retrieved from https://www.unicef-irc.org/publications/754-comparative-case-studies-methodological-briefs-impact-evaluation-no-9.html

Gugler, K., Rammerstorfer, M., & Schmitt, S. (2013). Ownership unbundling and investment in electricity markets — A cross country study. *Energy Economics, 40,* 702–713. https://doi.org/10.1016/j.eneco.2013.08.022

Handelsblatt (2007, February 2). Essent-Nuon merger costs German jobs [Essent-Nuon-Fusion kostet deutsche Arbeitsplätze]. Retrieved from https://www.handelsblatt.com/unternehmen/industrie/zusammenschluss-komplett-essent-nuon-fusion-kostet-deutsche-arbeitsplaetze/2764172.html

Henisz, W. J. (2000). The Institutional Environment for Economic Growth. *Economics and Politics, 12*(1), 1-31. https://doi.org/10.1111/1468-0343.00066

Hesseling D.& Sari M. (2007). The introduction of quality regulation of electricity distribution in the Netherlands. In Roggenkamp & Hammer (Eds.) *European Energy Law Report III* 2006, chapter 7 (pp. 127 – 145). Intersentia Ltd: Cambridge

Hogan & Hartson Raue (2007, August 28). *Separation of ownership of electricity and gas supply networks from other areas of energy supply* [Entflechtung des Eigentums an Elektrizitäts- und Gasversorgungsnetzen von anderen Bereichen der Energieversorgung], Retrieved from https://www.vzbv.de/sites/default/files/downloads/gutachten_entflechtung_hammerst ein_08_2007.pdf

International Financial Law Review. (2000). Liberalization of the Dutch Electricity and Gas Industries Retrieved from https://www.iflr.com/Article/2027656/The-Netherlands-LIBERALIZATION-OF-THE-DUTCH-ELECTRICITY-AND-GAS-INDUSTRIES.html

International Law Office. (2017). DSO unbundling – where are we now? Retrieved from https://www.internationallawoffice.com/Newsletters/Energy-Natural-Resources/Netherlands/Stek-Advocaten-BV/DSO-unbundling-where-are-we-now#

Jordana, J., Levi-Faur, D., & Puig, I. (2006). The Limits of Europeanization: Regulatory Reforms in the Spanish and Portuguese Telecommunications and Electricity Sectors. Governance, 19(3), 437-464. https://doi.org/10.1111/j.1468-0491.2006.00325.x

Joskow, P. L. (1996). *Introducing Competition into Regulated Network Industries.* Hierarchies to Markets in Electricity. Industrial and Corporate Change, 5(2), 341–382

Joskow, P. L., & Tirole, J. (2000). Transmission Rights and Market Power on Electric Power Networks. *The RAND Journal of Economics, 31*(3), 450. https://doi.org/10.2307/2600996

Kleinwächter, K. (2012): *Incentive regulation in the electricity industry in Germany. Positions of governmental and private actors [Die Anreizregulierung in der Elektrizitätswirtschaft Deutschlands. Positionen der staatlichen sowie privaten Akteure].* Potsdam: Universitätsverlag Potsdam

Knill, C., & Lenschow, A. (1998). Coping with Europe: the impact of British and German administrations on the implementation of EU environmental policy. *Journal of European Public Policy, 5*(4), 595–614. https://doi.org/10.1080/13501769880000041

Levi-Faur, D. (2004). On the "Net Impact" of Europeanization. *Comparative Political Studies, 37*(1), 3–29. https://doi.org/10.1177/0010414003260121

Lindemann, H. (2015). *Regulatory Objectives and the Intensity of Unbundling in Electricity Markets.* USAEE Working Paper No.15–200. Retrieved from https://ideas.repec.org/p/han/dpaper/dp-544.html

Looijestijn-Clearie, A. (2014). Breaking up Is Hard to Do: Dutch Unbundling Legislation and the Free Movement of Capital. *European Business Organization Law Review, 15*(3), 337-355. https://doi.org/10.1017/S1566752914001165

Mastenbroek, E., & Kaeding, M. (2006). Europeanization: Beyond the Goodness of Fit: Domestic Politics in the Forefront. *Comparative European Politics*, 4(4), 331–354

Meletiou, A., Cambini, C., & Masera, M. (2018). Regulatory and ownership determinants of unbundling regime choice for European electricity transmission utilities. *Utilities Policy*, 50, 13–25

Michaels, R. J. (2004). Vertical Integration and the Restructuring of the U.S. Electricity Industry. *SSRN Electronic Journal.* Advance online publication. https://doi.org/10.2139/ssrn.595565

Möllinger, Claus (2009): Eigentumsrechtliche Entflechtung der Übertragungsnetze unter besonderer Berücksichtigung des 3. Binnenmarktpaketes für Energie. Frankfurt, M., Berlin, Bern, Bruxelles, New York, NY, Oxford, Wien: Lang (Mannheimer Beiträge zum öffentlichen Recht und Steuerrecht, Bd. 31).

Monstadt, J., & Scheiner, S. (2016). Die Bundesländer in der nationalen Energie- und Klimapolitik: Räumliche Verteilungswirkungen und föderale Politikgestaltung der Energiewende. [The federal states in national energy and climate policy: spatial distribution effects and federal policy making for the energy transition]. *Raumforschung Und Raumordnung, 74*(3), 179–197. https://doi.org/10.1007/s13147-016-0395-6

Mulder, M., & Shestalova, V. (2006). Costs and Benefits of Vertical Separation of the Energy Distribution Industry: The Dutch Case. Competition and Regulation in Network Industries, 1(2), 197–230. https://doi.org/10.1177/178359170600100205

North, D. C. (1990). *Institutions, Institutional Change and Economic Performance.* Cambridge University Press. https://doi.org/10.1017/CBO9780511808678

OECD (1998). *Country Studies: Netherlands -Regulatory Reform in the Electricity Industry,* Retrieved from https://www.oecd.org/regreform/sectors/2497395.pdf

Pelkmans, J., & Luchetta, G (2013). *"Enjoying a single market for network industries",* Studies & Reports No. 95, Notre Europe, Paris, Retrieved from https://institutdelors.eu/wp-content/uploads/2018/01/singlemarketnetworkindustries-pelkmansluchetta-ne-jdi-feb13.pdf

Pierson, P. (2000). Increasing Returns, Path Dependence, and the Study of Politics. *American Political Science Review*, *94*(2), 251–267. https://doi.org/10.2307/2586011

Pollitt, M. (2007). The arguments for and against ownership unbundling of energy transmission networks. *Energy Policy*, *36*(2), 704–713. https://doi.org/10.1016/j.enpol.2007.10.011

Rosenberger, W. (2012). Is France taking back the EnBW network? [Holt sich Frankreich EnBW-Netz zurück?] *Stuttgarter Nachrichten*. Retrieved from https://www.stuttgarter-nachrichten.de/inhalt.geruechte-ueber-verkauf-holt-sich-frankreich-enbw-netz-zurueck.1c91217b-8a23-4b9d-88d6-2fc19b8421b1.html

Ruxton, G. D., & Neuhäuser, M. (2013). Review of alternative approaches to calculation of a confidence interval for the odds ratio of a 2 × 2 contingency table. *Methods in Ecology and Evolution*, *4*(1), 9–13. https://doi.org/10.1111/j.2041-210x.2012.00250.x

Scharpf, F.W. (2011). 'The JDT model: contexts and extensions', in G. Falkner (ed.), The EU's Decision Traps: Comparing policies, Oxford: Oxford University Press.

Sprent, P. (2011). Fisher Exact Test. In: M. Lovric (Ed.), *International Encyclopedia of Statistical Science* (pp. 524–525). Berlin: Springer.

Tsebelis, G. (1995). Decision Making in Political Systems: Veto Players in Presidentialism, Parliamentarism, Multicameralism and Multipartyism. *British Journal of Political Science*, *25*(3), 289–325. https://doi.org/10.1017/S0007123400007225

van Baal, P. A., & Finger, M. (2019). The Effect of European Integration on Swiss Energy Policy and Governance. *Politics and Governance*, *7*(1), 6. https://doi.org/10.17645/pag.v7i1.1780

van Binsbergen, J. H., & Marx, L. M. (2007). Exploring Relations Between Decision Analysis and Game Theory. *Decision Analysis*, *4*(1), 32–40. https://doi.org/10.1287/deca.1070.0084

van Damme, E. E.C. (2005). Liberalizing the Dutch Electricity Market: 1998-2004. *SSRN Electronic Journal*. Advance online publication. https://doi.org/10.2139/ssrn.869728

van Koten, S., & Ortmann, A. (2008). The unbundling regime for electricity utilities in the EU: A case of legislative and regulatory capture? *Energy Economics*, *30*(6), 3128–3140. https://doi.org/10.1016/j.eneco.2008.07.002

Verbund AG (2012), Annual Report 2011 [Geschäftsbericht 2011]. Retrieved from https://www.verbund.com/de-de/ueber-verbund/investor-relations/finanzpublikationen

WWU Münster. (2007). *Ownership Unbundling - Hoffnungen und Risiken. [Ownerhship Unbundling –*
Hopes and Risks], Retrieved from http://www.wiwi.uni-
muenster.de/06/nd/events/sonstige-events/ownership-unbundling-hoffnungen-und-
risiken/zusammenfassung/

Zachmann, G. (2007), *A Markov Switching Model of the Merit Order to Compare British and German Price*
Formation, DIW Discussion Papers 714

Zeit Online. (2009). Billion dollar deal in the electricity industry [Milliardendeal in der
Strombranche]. Retrieved from https://www.zeit.de/online/2009/09/vattenfall-
nuon

YOUR KNOWLEDGE HAS VALUE

- We will publish your bachelor's and
  master's thesis, essays and papers

- Your own eBook and book -
  sold worldwide in all relevant shops

- Earn money with each sale

Upload your text at www.GRIN.com
and publish for free